Our Exciting Earth!

LAKES

By Tanner Billings

Gareth Stevens Publishing

Please visit our website, www.garethstevens.com.
For a free color catalog of all our high-quality books, call toll free 1-800-542-2595 or fax 1-877-542-2596.

Library of Congress Cataloging-in-Publication Data
Names: Billings, Tanner, author.
Title: Lakes / Tanner Billings.
Description: New York : Gareth Stevens Publishing, 2023. | Series: Our exciting Earth! | Includes index.
Identifiers: LCCN 2021043387 | ISBN 9781538275689 (set) | ISBN 9781538275696 (library binding) | ISBN 9781538275672 (paperback) | ISBN 9781538275702 (ebook)
Subjects: LCSH: Lakes–Juvenile literature. | Lake ecology–Juvenile literature.
Classification: LCC GB1603.8 .B55 2023 | DDC 551.48/2–dc23/eng/20211008
LC record available at https://lccn.loc.gov/2021043387

First Edition

Published in 2023 by
Gareth Stevens Publishing
29 East 21st Street
New York, NY 10010

Editor: Kate Mikoley
Designer: Tanya Dellaccio

Photo credits: Cover Yevhenii Chulovskyi/Shutterstock.com; p. 5 Alicia Garcia Benito/Shutterstock.com; p. 7 Craig Sterken/Shutterstock.com; p. 9 Sergey Novikov/Shutterstock.com; p. 11 bikemp/Shutterstock.com; p. 13 John Brueske/Shutterstock.com; p. 15 Parichat Snguanwongwan/Shutterstock.com; p. 17 Katvic/Shutterstock.com; p. 19 Spiroview Inc/Shutterstock.com; p. 21 Suzanne Tucker/Shutterstock.com; p. 23 Zivica Kerkez/Shutterstock.com.

Printed in the United States of America

CPSIA compliance information: Batch #CSGS23: For further information contact Gareth Stevens, New York, New York at 1-800-542-2595.

Contents

A lake is a big body of water. It has land around it.

Most have fresh water.
This means no salt.

Some lakes have salt.
Utah has one.

Rivers go into lakes.

Some lakes are
made by ice.

Some lake water
is from rain.
Some is from the ground!

Lakes can be deep.
Russia has the deepest one.

North America has
the Great Lakes.
There are five.

You can swim in a lake.
You can ride a boat too!

Let's go to a lake!

Words to Know

river

water

Index